INSPECTION AND OTHER STRATEGIES FOR ASSURING QUALITY IN GOVERNMENT CONSTRUCTION

COMMITTEE ON INSPECTION FOR QUALITY CONTROL ON
FEDERAL CONSTRUCTION PROJECTS
BUILDING RESEARCH BOARD
COMMISSION ON ENGINEERING AND TECHNICAL
SYSTEMS
NATIONAL RESEARCH COUNCIL

William B. Ledbetter
Andrew C. Lemer
Editors

NATIONAL ACADEMY PRESS
Washington, D.C. 1991

NOTICE: The project that is the subject of this report was approved by the Governing Board of the National Research Council, whose members are drawn from the councils of the National Academy of Sciences, the National Academy of Engineering, and the Institute of Medicine. The members of the committee responsible for the report were chosen for their special competences and with regard for appropriate balance.

This report has been reviewed by a group other than the authors according to procedures approved by a Report Review Committee consisting of members of the National Academy of Sciences, the National Academy of Engineering, and the Institute of Medicine.

This report was prepared as part of the technical program of the Federal Construction Council (FCC). The FCC is a continuing activity of the Building Research Board, which is a unit of the Commission on Engineering and Technical Systems of the National Research Council. The purpose of the FCC is to promote cooperation among federal construction agencies and between such agencies and other elements of the building community in addressing technical issues of mutual concern. The FCC program is supported by 14 federal agencies: the Department of the Air Force, the Department of the Army, the Department of Commerce, the Department of Energy, the Department of the Navy, the Department of State, the General Services Administration, the National Aeronautics and Space Administration, the National Endowment for the Arts, the National Science Foundation, the U.S. Postal Service, the U.S. Public Health Service, the Smithsonian Institution, and the Department of Veterans Affairs.

Funding for the FCC program was provided through the following agreements between the indicated federal agency and the National Academy of Sciences: Department of State Contract No. 1030-621218; National Science Foundation Grant No. MSM-8902669, under master agreement 8618641; and U.S. Postal Service grant, unnumbered.

Library of Congress Catalog Card No. 91-62333

International Standard Book Number 0-309-04547-9

Additional copies of this report are available from: National Academy Press 2101 Constitution Avenue, NW Washington, DC 20418

S 409

Printed in the United States of America

COMMITTEE ON INSPECTION FOR QUALITY CONTROL ON FEDERAL CONSTRUCTION PROJECTS

Chairman

WILLIAM B. LEDBETTER, PE, Consultant, Pendleton, S.C.

Members

MERRILL R. COTTEN, Chief, Construction Management, De Leuw, Cather & Company, Chicago, Illinois

F. H. (BUD) GRIFFIS, Professor of Civil Engineering, Columbia University, New York, New York

C. RAYMOND HAYS, Director Safety/Quality, RUST International, Birmingham, Alabama

ARTHUR T. KORNBLUT, Kornblut & Sokolove, Cheve Chase, Maryland

JOHN W. MORRIS, USA Retired, Engineer Advisor, Zorc, Rissetto, Weaver & Rosen, Washington, D.C.

DONALD W. PFEIFER, Vice President, Wiss, Janney, Elstner Associates, Northbrook, Illinois

ROBERT ULREY, Manager of Technical Services, Houston Lighting and Power Company, Houston, Texas

DAVID J. WICKERSHEIMER, President, Wickersheimer Engineers, Inc., Champaign, Illinois

PAUL ZIA, Distinguished University Professor, Civil Engineering Department, North Carolina State University, Raleigh

Federal Construction Counsil Liaison Representatives

KNOX BURCHETT, Office of Foreign Building Operations, Department of State
WILLIAM DUNCAN, P.E., Division of Health Facilities Planning, Public Health Service
JAMES D. DURHAM, Office of Environmental Health and Engineering, Indian Health Service
ROBERT EIDSON, Office of Facilities, Department of Veterans Affairs
IRVIN HAMBURGER, P.E./C.P.E., Office of the Air Force Civil Engineer
JAMES R. JONES, Directorate of Military Programs, U.S. Army Corps of Engineers
WILLIAM J. KEHEW, P.E., Chicago Operations Office, U.S. Department of Energy
ALEX LINCOLN, Public Buildings Service, General Services Administration
MIGUEL LOPEZ, Naval Facilities Engineering Command, Department of the Navy
WILLIAM POUNDS, Office of Foreign Buildings Operations, Department of State

Public Facilities Council Liaison Representatives

ROBERT KELLER, Division of Design and Construction, Office of Administration, State of Missouri
FRANCIS W. KUCHTA, Department of General Services, State of Maryland

Project Staff

ANDREW C. LEMER, Director
PETER H. SMEALLIE, Senior Program Officer
PATRICIA M. WHOLEY, Staff Associate
SUZETTE CODY, Project Assistant
JOANN V. CURRY, Senior Secretary

BUILDING RESEARCH BOARD (1990–1991)

HAROLD J. PARMELEE, President, Turner Construction Company, New York, New York

LESLIE E. ROBERTSON, Director, Design and Construction, Leslie E. Robertson Associates, New York, New York

JAMES E. WOODS, William E. Jamerson Professor of Building Construction, College of Architecture and Urban Studies, Virginia Polytechnic Institute and State University, Blacksburg, Virginia

APRIL L. YOUNG, CRA Coordinator, First American Metro Corporation, McLean, Virginia

Staff

ANDREW C. LEMER, Director
HENRY A. BORGER, Executive Secretary, Federal Construction Council
PETER H. SMEALLIE, Senior Program Officer (to June 1990)
PATRICIA M. WHOLEY, Staff Associate
SUZETTE CODY, Project Assistant
JOANN V. CURRY, Senior Secretary
LENA B. GRAYSON, Program Assistant

PREFACE

Quality is not an act. It is a habit.
Aristotle

Much has been said and written in recent years about the nation's declining commitment to the principles of quality that marked our rise as the world's leading industrial power and about our need as a nation to renew that commitment. President Bush has stated that ". . . our work to build quality products will be a crucial link to the long-term success of the United States in the global marketplace."

The U. S. construction industry has not been immune to the problem. Our share of world construction markets has shrunk. Foreign competitors are gaining increasing strength in our domestic markets. Many people in the industry point to quality—in our materials and equipment, our design and workmanship, in the private and public sectors—as significant causes.

When the agencies of the Federal Construction Council requested the Building Research Board to advise them on construction inspection practices that would help to improve the quality of their constructed facilities, they referred only to the tip of this massive iceberg of a problem confronting the entire industry. We and the committee invited to provide this advice found it difficult to restrict our attention to this one small aspect of the major challenge of moving the industry back to building quality facilities.

However, we recognize that habits are persistent and, once formed badly, can be changed only through constant attention and perseverance. Inspection is an important and long-accepted means to achieving quality and one that is well suited to the task of changing bad habits. We hope that our work and this report will contribute to a change in habits and thereby to the broader goals of quality to which we all must aspire.

William B. Ledbetter, *Chairman*

Committee on Inspection for Quality Control on Federal Construction Projects

Andrew C. Lemer, *Director*

Building Research Board

EXECUTIVE SUMMARY

Quality facilities enhance the safety, productivity, and environment of the people and economic activities they serve, and are achieved through a process of planning, design, construction, operation, and maintenance that must work effectively at all stages. In construction, quality is obtained through conformance to adequately developed requirements, requirements clearly stated to set forth the characteristics which the constructed facility must have to serve its users well.

An essential precondition for assuring construction quality is getting the requirements right and stating them clearly and accurately in the drawings and specifications to be followed by the constructor. Construction costs that exceed design estimates are an important indicator of failure to achieve quality. If the requirements are right and the estimating effective, then cost growth is attributed to construction problems, but in practice the growth is largely attributable to design inadequacies. However, achieving quality in construction remains a substantial challenge after planning and design have produced an appropriate and clear statement of requirements.

The constructor has a responsibility to control quality and conform to requirements by managing labor, equipment, and materials to produce a facility that meets these requirements. The purchasers of construction—the owner and users of the facility—seek to assure that quality by participating in the monitoring of the constructor's quality control activities. Inspection—the specific examination, testing, and overall appraisal of a process, product or service to ascertain if it conforms to established requirements—is an essential tool for construction quality control and assurance.

Federal agencies follow a variety of practices in using inspection to assure quality. Some agencies assign inspection responsibilities to their own staff. Others depend on the design architect or engineer or an independent firm to conduct inspection. Some agencies rely almost entirely on the constructor's quality control system and conduct few inspections of their own. All of these various practices have produced comparable and generally adequate construction quality, but the U. S. Army Corps of Engineers' contractor evaluation procedures are exemplary. In general, quality could be improved by reducing this unnecessary variety in practices, by making inspection more efficient, and by avoiding specific practices that foster conflict among owners, designers, and constructors.

The U. S. construction industry and its customers have much to gain from the growing interest in teamwork and long term commitment to quality reflected in the philosophy and practices presented under such titles as Total Quality Management. Owners and users working to enhance their assurance of quality in construction should act within the context of this philosophy and practice. Federal agencies in particular can enhance the quality of their construction by taking action on the following recommendations:

1. Work to improve the agency's ability to develop quality definitions, facility programs, plans, budgets, guide criteria,

design drawings, and specifications that convey correct requirements in a clear manner to the constructor.

2. Avoid adversarial design and construction management practices and adopt defined programs to foster teamwork among users, design and construction managers, designers, and constructors.
3. Assure that drawings and specifications are complete, clear, and consistent.
4. Avoid overly frequent or detailed inspections that do not contribute directly to assuring final quality.
5. Join with other agencies in specific programs to share information and centralize selected inspection activities, for example by participating in the Army's Construction Contractor Appraisal Support System.
6. Work with designers and users to develop integrated inspection plans for all construction projects, and then with constructors to assure that the plans are effectively executed.
7. Fund research and demonstration activities required to develop new inspection and other quality assurance technologies.
8. Adopt systems to measure explicitly the agency's quality management efforts and to relate those efforts to the costs of replacement and repair of faulty construction or of productivity lost through acceptance of faulty construction.
9. Assert, at senior administrative levels, each agency's commitment to quality in its constructed facilities and establish definite programs for making the commitment effective and lasting.

CONTENTS

INSPECTION AND OTHER STRATEGIES FOR ASSURING QUALITY IN GOVERNMENT CONSTRUCTION

1

INTRODUCTION

Federal agencies typically spend some $4 billion to $5 billion annually for construction of new and substantially renovated buildings, and total spending for construction of all federal facilities in 1989 was roughly $12.7 billion (Department of Commerce, 1990). State and local governments together spend approximately another $25 billion for buildings and $71.6 billion overall each year. The products of this spending are public assets that serve a wide range of private and public purposes and have pervasive influence on the productivity and quality of life of everyone. Assuring that these products meet the highest possible quality standards is a major challenge. Purchasers of construction in the private sector face this challenge as well, and some observers suggest the challenge is being poorly met in both private and public sectors.

The Business Roundtable's Construction Industry Cost Effectiveness (CICE) Project noted quality problems influencing the declining productivity of the nation's construction industries.[1] That group's 1983 report cited evidence of a rapid growth of "disputes involving liability, negligence, claims for errors and omissions, and governmental citations," and called for better quality control.

In 1984, a workshop of nearly 100 representatives of the design and construction industry, convened in Chicago under the auspices of the American Society of Civil Engineers (ASCE), agreed that "accidents, design flaws, cost overruns, and other similar problems were occurring at a serious rate."[2] That group's discussions motivated ASCE to undertake development of a comprehensive guide to quality in design and construction.

Regardless of current quality, facilities owners and the construction industry share a common concern for the quality of our buildings and other constructed facilities. The sponsors of the Federal Construction Council (FCC),[3] government agencies responsible for managing the public's assets, feel this concern most keenly. Managers of some of these agencies find that budgetary constraints and other technical and administrative forces increasingly threaten the effectiveness of their inspection programs and thereby pose serious impediments to achieving quality. Others have delegated the inspection task to contractors and find themselves accused of abandoning all hope of achieving quality by asking foxes to guard the henhouse.

The FCC agencies asked the Building Research Board (BRB) of the National Research Council to undertake a study of inspection as a means to control quality in construction. The BRB selected a committee of professionals with broad expertise

[1] The Business Roundtable, 1983. More Construction for the Money, Summary Report of the Construction Industry Cost Effectiveness Project.

[2] American Society of Civil Engineers, 1988. Quality in the Constructed Project: A Guideline for Owners, Designers and Constructors, Volume 1, preliminary edition for trial use and comment.

[3] Sixteen federal government agencies with major interests in building and facilities research, construction, operation, and maintenance comprise the Federal Construction Council. These agencies had a combined responsibility for facilities-related budgets in FY 1989 exceeding $17 billion.

and extensive experience to undertake this study.[4] This document is a report of that committee's deliberations.

SCOPE OF THE COMMITTEE'S DELIBERATIONS

The BRB's committee examined the factors that affect the need for inspection, and the extent and nature of inspection during construction, with the goal of recommending techniques for enhancing the value of inspection that will achieve more cost-effective construction of federal projects. Most federal facilities are constructed under fixed price contracts awarded on the basis of open competitive bidding. Agencies use detailed design criteria and construction specifications to describe the characteristics of materials and workmanship required, and use inspection to monitor contractor compliance.

The committee met several times during a period of about one year, and heard testimony of federal agency representatives and experts in the private sector and academia. Early in their deliberations, the committee agreed that construction quality is inextricably related to design quality, and that inspection is only one of a number of methods for assuring quality in the constructed facility. More importantly, the committee found that many of the problems perceived with construction quality in the United States today are beyond the reach of inspection. While the committee's work was focused on inspection, the committee's report unavoidably touches on design and other strategies for achieving quality.

In particular, the committee asserted that in their judgment *major problems of quality in* U.S. *construction today begin in planning and design.* Construction contractors are expected to deliver facilities that conform to requirements presented in drawings and specifications prepared by planners and designers. If these drawings and specifications are not accurate, complete,

[4] Biographical sketches of the committee members are presented in Appendix A.

and clearly presented, or if they describe a facility unlikely to meet the needs of owner and users, even the most careful and conscientious construction will not deliver quality. Cost growth in construction—costs greater than were estimated in design—is one indicator of lost quality.

If the design is effective in getting the requirements right and estimating accurately, then cost growth is attributable to construction problems. However, this is not often the case. One study in the private sector involving the construction of nine fast-track[5] industrial projects revealed that the cost of repair or replacement (rework), an average of more than 12 percent of the total installed project costs, was attributable primarily to design errors (25 percent) and owner and designer changes (54 percent), and only 17 percent to construction errors.[6] These rework costs were borne by the owner in the form of contract modifications or change orders.

In the federal sector, design is generally completed prior to issuance of construction contracts. A study of projects constructed for the U. S. Army Corps of Engineers and the Naval Facilities Engineering Command (NAVFAC) found average cost growth of approximately 6 percent attributable primarily to design error (50 to 65 percent) and owner changes (23 to 35 percent).[7] More recent data furnished the committee by the Corps and NAVFAC indicate that similar cost growth is still

[5] "Fast-track" is the term used for projects whose construction begins before all design is completed. Fast-track procedures are used to reduce the time between start of design and construction completion.

[6] The Quality Management Task Force, 1989. Cost of Quality Deviations, CII Publication 10-1, The Construction Industry Institute, The University of Texas at Austin.

[7] Building Research Board, 1986. Construction Contract Modifications: Comparing the Experiences of Federal Agencies with Other Owners. National Academy Press, Washington, DC. Some agencies report that owner changes are much higher.

being experienced, and earlier studies show that the scale of the problem is not new.[8]

ORGANIZATION OF THE REPORT

While the committee focused on government—particularly at the federal level—such cases illustrate that both private and public sectors share the problem of assuring construction quality. The committee's deliberations thus considered both sectors, and this report is meant to have a broad bearing on the construction industry as a whole.

The following pages summarize the committee's discussions of construction quality and current practices for its assurance, highlighting the role of inspection. Chapter 2 presents the underlying principles and definitions the committee adopted for their discussions. Chapter 3 describes inspection strategies of various federal agencies and the private sector, highlighting some of the latter that are likely, in the committee's assessment, to be particularly effective for assuring quality in federal construction. Chapter 4 considers the limitations of inspection and presents selected alternative strategies to enhance inspection for assuring quality. Chapter 5 summarizes the committee's specific recommendations for achieving government construction quality. Appendices present supplemental information on topics introduced in these chapters.

Quality in construction occurs through a complex interaction of many participants in the facilities development process. The committee's recommendations are aimed primarily at agency managers, but address design and construction professionals, educators, and policy makers as well. The committee agreed that quality in construction is assured only when there is a commitment to quality throughout planning, design, and con

[8] Building Research Board, 1986. Supervision and Inspection of Federal Construction, Federal Construction Council Technical Report No. 54, National Academy Press.

struction, and that quality facilities require that this commitment continue through operations and maintenance. Quality facilities that meet and exceed expectations —enhancing our safety, productivity, and overall quality of life—are the real goal of the committee's deliberations.

2

QUALITY AND PRACTICES FOR ITS ASSURANCE

The term quality, particularly when applied to facilities, has no single generally accepted meaning. For many people, it is a characteristic of an object. For others, it has something to do with actions taken to achieve an object with certain characteristics. Quality is a value-laden term that depends on one's point of view.

According to the dictionary, *quality* means "a degree of excellence . . . superiority in kind."[9] The authors of ASCE's manual defined quality as "the totality of features, attributes, and characteristics . . . that bear on . . . ability to satisfy a given need: fitness for purpose . . . meeting the requirements."[10] The American Society for Quality Control termed quality "a

[9] Mish, Frederick C., ed. 1985. Webster's Ninth New Collegiate Dictionary, Merriam-Webster, Inc., Springfield, MA. A glossary of terms is presented in Appendix B.

[10] ASCE, op. cit., p. 17.

systematic approach to the search for excellence." Another committee of the BRB found it impossible to devise a concise, complete, and generally acceptable definition of design quality and for working purposes referred to quality buildings as those "whose characteristics create an environment where the occupant or user can accomplish his purpose effectively, efficiently, and comfortably."[11]

DEFINING CONSTRUCTION QUALITY, ASSURANCE, AND CONTROL[12]

Within the limited context of the design and construction stage of a facility, quality can be more readily defined, and the committee accepted a definition that *quality is conformance to adequately developed requirements.* This definition indicates that a quality constructed facility will result provided that several conditions are met:

1. The contract documents comprise a clear, complete, and accurate description of the facility to be constructed, correctly conveying the intent of the owner regarding the characteristics of a facility needed to serve his or her purposes.
2. The contract documents define a constructed facility considered acceptable under applicable regulatory codes and standards of professional practice, in terms of its reliability, the ease with which maintenance and repairs can be performed, the durability of its materials and operating systems, and the life safety afforded its users.
3. The facility is constructed in accordance with those documents.

---

[11] Building Research Board, 1989. Improving the Design Quality of Federal Buildings, National Academy Press, Washington, DC.

[12] Refer to Appendix B for definition of terms.

In other words, quality in construction dictates both that the requirements are met and that the requirements be right. The committee's deliberations focused on the first aspect of quality—meeting the requirements—but unavoidably addressed some of the concerns for getting the requirements right in the planning and design that precede construction.

Purchasers of construction—building owners, developers, and their representatives—use a variety of methods to assure that the construction and resulting facility meet their requirements. Planned and systematically organized, these methods comprise a *quality assurance* (QA) program.

Providers of construction services and manufacturers of construction products and building systems also use a variety of methods to make sure that their products—including entire facilities—meet the requirements. *Quality control* (QC) is accomplished through these various methods, which may be formally organized into a control system or informally undertaken by managers and workers.

One other term is important: *quality management.* Quality management is the process of optimization of quality activities, and includes problem prevention and quality appraisal activities. As such, it involves both quality assurance and quality control.

There are no generally accepted definitions of the terms quality assurance and quality control, and the two are used interchangeably by many people. The committee adopted a distinction that quality control is what a construction contractor does to determine that the products of his or her work—the completed facility—conform to the requirements stated in contract documents. Quality assurance is what the purchaser of construction does to determine that contractor's quality control system is functioning adequately and that the product consequently will meet the purchaser's needs.[13]

[13] This statement presumes that the requirements given to the constructor are an adequate statement of the purchaser's needs. When this is not true, quality assurance efforts may involve reprogramming and redesign, and loss of quality.

Just as there are no generally accepted definitions for quality assurance and control, there is no general understanding of what the costs of these activities are or should be. Elements of QA and QC are performed as part of planning and design, and even in the construction stage are not all clearly distinguished in the cost accounting. Even when QA and QC activities are explicitly identified, practices vary so much among projects and contexts that no reliable base of information exists for making general judgments about costs in general construction.

INSPECTION FOR QUALITY CONTROL AND ASSURANCE

QA and QC systems include management reviews, on-site surveillance, and tests. *Inspection*—specific examination, testing, and overall appraisal of a process, product or service to ascertain if it conforms to established requirements—is standard practice for quality assurance and control in all major construction. However, the degree to which inspection can be successful as an assurance and control method is limited by the established requirements. If the established requirements are not right, inspection cannot make them right.

Most construction contracts—particularly those issued by public agencies—specify that all work is subject to inspection by the owner or the engineer or architect representing the owner. The contract specifications may then enumerate a list of specific inspections that will be required during construction. These inspections are called *controlled inspections* . Contractors offering to construct facilities know what inspections will be required, before they enter into a contract agreement, because they are listed in the specifications. Controlled inspections include examination and approval of products prior to their installation (for instance precast concrete structural members) and activities incident to the construction (such as dewatering and wastewater discharge). The contract documents may authorize the architect or engineer to review and approve the contractor's methods and require the contractor to provide certifications from an approved testing laboratory that materials meet specified standards. Inspection and in-place testing of completed work on critical components (such as chilled or hot water lines, or welds on jet fuel lines) are usually included among specified controlled inspections.

The owner may elect to conduct additional inspections, termed *discretionary inspections*, to further enhance quality assurance. The owner or the owner's representative may choose to conduct these discretionary inspections depending on the circumstances of the construction project. The location, complexity and criticality of the project, the availability of trained inspectors, the background and experience of the owner's representative, and the tradition in an agency or geographic area influence the degree to which discretionary inspections are performed.

Inspection is generally given more emphasis as a quality assurance tool when there has been a demonstrated deficiency in a contractor's performance. Federal agencies cannot depend only on the reputation of the contractor as a basis for disqualification or for scheduling inspections, but may adjust their quality assurance activities in response to demonstrated performance. (This "reputation" factor is utilized routinely in the private sector.) However, unwarranted emphasis on inspection—particularly when on-site representatives of the owner are responsible, as is frequently the case in government construction—tends to foster unproductive adversarial relationships between the contractor and the owner. The appropriate level of inspection must be adequate to assure quality but not so much as to reduce productivity.

THE FEDERAL ACQUISITION REGULATIONS

Review of each agency's method of obtaining conformance with the design documents by the construction contractor reveals differences in the organization for and practice of inspection among these agencies. In large measure, the variations are the result of or are driven by such factors as an agency's historical experience, institutional priorities for use of manpower and dollars, types of project, and sources of funds. In spite of these variations there remains a thread of consistency generated by the need to conform to the Federal Acquisition Regulation (FAR).

The Federal Acquisition Regulation (FAR) establishes the core procedures used by the FCC sponsors and other federal Executive-branch agencies in their acquisition of supplies and services with Congressionally appropriated funds. The FAR system, developed in accordance with the requirements of the Federal Procurement Policy Act of 1974 and subsequent amendments, is issued under the joint authorities of the Administrator of General Services, the Secretary of Defense, and the Administrator for the National Aeronautics and Space Administration, under the broad policy guidelines of the Administrator for Federal Procurement Policy.

The FAR replaced and consolidated the older Federal Procurement Regulation (FPR) and Armed Services Procurement Regulation (ASPR). However, supplemental regulations conforming to the FAR are issued by the individual agencies. The FAR precludes agency acquisition regulations that unnecessarily repeat, paraphrase, or otherwise restate the FAR and it limits agency acquisition regulations to those necessary to implement FAR policies and procedures within an agency. However, individual agencies determine what repetition and restatement are necessary to the agency's mission. Appendix C lists those portions of the FAR that most affect construction projects and construction contract administration.

The FAR does not specifically define the terms quality, quality control, or quality assurance, but does include definitions and guidance relevant to construction quality. *Contract quality requirements* are those technical statements in the construction contract that relate to the quality of the product or service and the inspection or other quality controls required of the contractor to assure conformance to contract requirements.

Government contract quality assurance is the various functions, including inspection, that the government performs to determine that the contractor has fulfilled contract obligations of quality and quantity. Inspection means examining and testing supplies or services, possibly including raw materials, components, and intermediate assemblies—as well as work-in-place—to determine their conformance to contract requirements.

Part 52 of the FAR, "Contract Clauses," requires agencies to include in their fixed-price[14] construction contracts a standard clause requiring contractors to "maintain an adequate inspection system and perform such inspections as will ensure that the work called for by this contract conforms to contract requirements" and to "maintain complete inspection records and make them available to the government." Contracts with a value below a defined small purchase amount[15] are exempted from the requirement, but agencies may still choose to include the clause. Most agencies have adopted much more elaborate sets of regulations and requirements for quality assurance.

AGENCY QUALITY MANAGEMENT PRACTICES

Within the common framework established by the FAR, individual federal agency QA and QC practices vary substantially as to where responsibilities are assigned and the formal components of the QA/QC program. The Department of Veterans Affairs (VA), for example, holds the general contractor responsible for quality control and VA resident engineers—assisted by the design firm and a government-hired testing laboratory—monitor and inspect to ensure contractor

[14] Most governmental construction in the United States is purchased under fixed-price arrangements, in which contractor and owner agree to a definite total amount to be paid for satisfactory completion of construction of a facility described by previously prepared drawings, specifications, and related contract documents.

[15] A small purchase is defined in the FAR as amounting to $25,000 or less.

compliance. The Army Corps of Engineers and Naval Facilities Engineering Command (NAVFAC) also leave QC generally to the contractor, but often require that the contractor perform specified inspections, and sometimes supplement their staff QA personnel with third party professionals to verify contractor compliance.

The Office of Foreign Buildings Operations of the U.S. Department of State (DoS), responsible for overseas embassy construction, has adopted an approach to quality management similar to the Corps. The construction contractor is required to provide an inspection system, and certain tests, inspections, and reports are specified. The DoS places on each project a staff project director to administer a program of testing and inspections for quality assurance. This program may often be quite stringent because of the unique problems associated with embassy security requirements and design standards and the challenges of construction in overseas locations.

The General Services Administration (GSA) assumes primary responsibility for quality assurance and control during the construction phase of its projects, while recognizing that its construction contractors support the QA/QC effort with their own internal quality control systems. To perform on-site quality control functions, GSA uses in-house staff, personnel provided by construction management or the design A/E consultants, or various combinations of these.

Some agencies supplement the FAR with additional regulations intended to respond to specific characteristics of their missions. The Army Corps of Engineers, for example, is regulated by the Department of Defense's FAR Supplement (DFARS), the Army FAR Supplement (AFARS) and the Engineer FAR Supplement (EFARS), in descending order of hierarchy. The Naval Facilities Engineering Command supplements the FAR and DFARS with the Navy Acquisition Procedure Supplement (NAPS). The Department of Veterans Affairs adds the Veterans Administration Acquisition Regulations (VAAR). These lower level regulations usually add more specific and restrictive language to the basic FAR requirements and must be read in conjunction with all higher level regulations.

The Department of Energy (DoE), whose construction of research, nuclear, and other unique facilities demands specialized oversight and regulatory controls, has developed its own detailed quality assurance requirements [16]. In most cases, DoE personnel are directly involved with QA activities. On small projects, inspection may be carried out by A/E personnel. On large projects, construction management contractors may perform inspections as part of the QA function.

Because its construction is administered primarily by the Army Corps of Engineers, the Air Force does not maintain its own construction inspection staff. The Air Force nevertheless deserves mention: leadership of the Air Force has asserted strongly its commitment to the quality of its facilities and staff professionals work with their designers and construction agency counterparts to assure that this quality is delivered. The ability of construction agencies to assure quality facilities is greatly enhanced by such strong user commitment.

In spite of the current variations in inspection practice by government agencies, there seems to be a consistent trend within the engineering and construction profession which leads the public agencies toward smaller and smaller forces to provide inspection oversight. This movement can progress only so far in public contracting because present laws preclude rewarding good performance by new, non-competitive construction contracts as is the case in private industry. This constraint may reduce the contractor's incentive for excellence because he or she must compete regardless of past quality or level of performance, but it should not reduce the contractor's pride in work well done or the need to build and retain a reputation for excellence. The federal agencies should foster and assist the growth of this professional attitude in those contractors that serve public programs.

[16] These are contained in DOE Orders 5700.6B, "Quality Assurance," 4700.1, "Project Management," and 6430.1A, "General Design Criteria."

FACTORS THAT INFLUENCE AGENCY PRACTICES

Agency liaison representatives to the committee suggested that the variations among QA practices of individual agencies are traceable to a number of factors that fall broadly into four categories: (1) budgetary, (2) agency programs, (3) project-specific, and (4) personnel-related factors.

Budgetary factors include both the amount of funding available for construction control as well as the allocation of construction funding to inspection and other functions. Recurring budget cutbacks tend to place particular pressures on quality management because the loss of quality is often difficult for the untrained eye to observe. Restrictions on funds use (i.e., ear-marking), imposed in the authorization and appropriation process or in top-level agency management decisions, may limit an agency's ability to pay for QA. Construction cost increases associated with changes in the requirements, termed "upgrading," is rarely matched by adequately increased allocations to QA activities needed to maintain quality oversight of the expanded project.

The particular types of facilities an agency develops and the criticality of these facilities to the agency's mission will influence the agency's inspection practices. When the risk of loss from non-conformance to the requirements is high, as in an embassy built by the Department of State or a strategic defense facility built by the Corps or NAVFAC, greater effort is devoted to insuring the requirements are met. Agencies that build for other agencies, such as the Corps working for the Air Force, must use inspection strategies that meet customer expectations, in a manner similar to the private sector.

Projects of large scale or complexity always require greater attention, and agencies that build such projects frequently are inclined to be more stringent in all of their QA activities. (This is true of non-federal agencies as well. See box.) Issues of national security, social sensitivity, and environmental impact contribute to complexity of certain projects, as will remote or inhospitable geographic locations.

QA AND QC AT KENNEDY AIRPORT'S RENEWAL

The Port Authority of New York and New Jersey (PANY/NJ) in its JFK2000 Redevelopment Program for Kennedy International Airport in New York City has adopted a policy that QC--the process of ensuring that proper materials and equipment are furnished and used, competent workmanship is provided, and timely services are performed in accordance with the contract requirement--is the contractual responsibility of the contractor. This responsibility includes providing inspection and inspection reporting, systems testing as required by the contract, providing survey control, preparing as-built drawings, maintaining inspection and systems testing documentation, including off-site quality control records such as manufacturer's certificates of compliance, and submitting copies of all contract documentation to the construction resident engineer. The contractor must submit a proposed QC program to the construction manager for review and approval prior to the start of work in the field.

The construction manager (CM) is responsible for quality assurance. The program manager is responsible for auditing the CM's administration of the QA/QC program. The Engineer of Record has some specific inspection requirements such as test pile measurements and inspections of each pile prior to concrete placement.

The PANY/NJ requires the CM to assign inspectors who have the primary day-to-day responsibility for confirming that the contractor's work is in accordance with the specifications, the approved QC program, and all applicable codes. Documentation that work has been performed satisfactorily includes laboratory and test results and inspection reporting. The inspector is responsible for a number of specified tasks:

- Confirming when and where routine testing will be required and arranging with the laboratory to have tests performed;
- Providing lab personnel with needed information;
- Witnessing all testing and verifying that requirements were followed;
- Arranging for specialist assistance for witnessing testing, as required;
- Recording all testing on a Daily Construction Report;
- Documenting all areas of nonconformance;
- Maintaining copies of test results, inspection reports, certification papers and permits;
- Verifying that testing devices are calibrated;
- Coordinating site activity;
- Visual inspection of all items not requiring laboratory testing;
- Preparing and maintaining inspection checklists.

On the other hand, repetitive facilities such as military barracks can involve substantial off-site fabrication that reduces on-site inspection requirements. Some agencies, such as the Department of State, are able to develop longer term relationships with contractors, such that these contractors become thoroughly familiar with agency requirements and the agency may, with confidence, reduce the level of effort in its QA program.

The continuity of progress of a particular project through design and construction influences the need for inspection as well. Delays and interruptions on a project can result in changed specifications and changed requirements that then necessitate greater QA effort. The urgency of completing construction leads too easily to neglect of QA activities that might otherwise have been undertaken.

Personnel-related factors may be the most significant determinants of agency QA practices. Agency personnel often believe that staffing levels are insufficient for the range of administrative responsibilities they face, and that quality assurance activities are frequently sacrificed. Because of budget cutbacks, low pay, normal retirements, and a general shortage of trained professionals, many government agencies have suffered losses of experienced staff to execute or supervise QA activities. Personnel regulations, compensation, and mobility requirements make it difficult for these agencies to attract and retain qualified professionals, which both increases training costs and reduces effectiveness of their QA activities. The task of inspection, in particular, is in some agencies viewed as unlikely to contribute to career advancement. [17]

[17] These various factors have led to increased use of consultants when budgets have permitted (Newman, 1989).

PERSPECTIVES FROM PRIVATE SECTOR QUALITY MANAGEMENT PRACTICES

The committee recognized that federal agencies face a number of unique conditions and responsibilities that shape their methods of quality assurance and quality control. The committee made a number of general observations on private sector practice that provide perspective for assessing federal agencies' inspection and other quality management activities.

One important difference between federal and private sector construction is that virtually all private design and construction is negotiated in some fashion, and repeat business with major clients depends on (1) cost management, (2) schedule management, and (3) quality management. Since repeat business is very important to contractors engaging in private sector work, contractors have a high stake in providing a quality product. Negotiation is used by government also, but competitive bidding is by far the more common practice.

A second important difference is that most private sector construction involves contractor pre-qualification. In the public sector, if a bidder can obtain a performance bond, he or she normally is entitled to bid.[18] The contracting officer determines that a low post-bidder is responsible before the contract is awarded. On private work, the owner often selects a single contractor, or a small group of contractors deemed to be qualified.

A third important difference is that the private sector is subject to legal regulation by local building codes. Most cities and municipalities require an inspection by local government personnel to assure conformance to code requirements. Designers and constructors for the private sector often rely on these codes and municipal inspection, especially with respect to electrical and plumbing systems, to assure the quality required. Upon a building's completion, most responsible local authorities issue an occupancy permit that certifies that a building permit

[18] The bond is furnished after acceptance of the bid.

was obtained at the start of construction, that electrical and plumbing inspections were made, and that the fire department has also inspected and approved the building.

Federal agencies are not subject to local codes and municipal inspection. The agencies use their own design criteria and must assure for themselves that these criteria are followed in design and construction.

In contrast to the federal government, the private sector encompasses organizations with a broad range of size and only the largest firms have building professionals on staff. Even these large firms, unless they maintain steady building programs, are unlikely to have sufficient staff capability to undertake substantial in-house quality management activities. Most private sector owners depend primarily on A/Es and other consultants to act as their QA agents and pay these agents for the service.

Very active builders, such as real estate developers and certain large manufacturers, might be more likely to undertake extensive QA programs but this is not often the case. Real estate developers, in particular, rarely are willing to pay the A/E firm additional fees (typically 3 to 4 percent of construction) for full field inspection, and quality assurance inspections during construction will be casual and infrequent. The QA tasks are often tailored to whether the items in question are exposed or hidden from view in the completed project. Items hidden from view include structural concrete and steel, underground utility lines, foundations, pilings, caissons, service lines inside walls or above ceilings, water lines, gas lines, telephone lines, electrical conduits, sprinkler systems, HVAC duct work, and plumbing. On the private sector construction site, these hidden items may receive less inspection attention than on public works.

On the other hand, many owners who are going to occupy as well as operate their buildings seek to assure the quality of the finished building by hiring an A/E firm or other qualified individuals to act as the owner's representative to supervise the construction. Independent testing laboratories will be used to perform the actual tests. The amount of testing ordered is usually determined by the level of in-house expertise and the owner's relationship with the contractor.

The quality management philosophy of much private commercial work is expressed in the common phrase "If it don't fit, make it." When errors or other problems occur, alterations are made until there is a workable fit. Rarely is work scrapped or demolished and then re-executed as shown on the drawings and specifications. Because time delays are costly to private sector builders and owners, there is great pressure for the work to proceed uninterrupted on the construction site. Government agencies are seldom charged interest costs and so are not exposed directly to the strictly financial costs of delay. (However, contractors may charge the government for government-caused delays.)

The A/E firm's role in private work inspection is similar to public sector work, except that procedures are generally less formal and there tends to be more practical interaction between the contractor staff and the owner's representatives.[19] A/E firms that work primarily for distinct groups or types of private clients (e.g., retail or office commercial, large residential, or light industrial) often must adjust their staffing and management practices to accommodate the particular style of QA that project type entails. The size of the A/E firm, often closely correlated with the scale of projects the firm designs, also influences how the QA function is carried out.

For example, low overhead is essential to the success of a small A/E office. Generally, there is no dedicated staff to handle quality management activities, such as quality assurance and inspection for quality control on projects. These tasks are performed by the same professional staff members that design, draft, and manage projects. Small, private clients question the value of the added expense of quality assurance, often assuming

[19] One sharp difference is that the A/E may be authorized to act as the owner's agent, a practice virtually unknown in federal agency construction.

that the contractor's quality control program is sufficient to assure conformance with construction documents.

Small A/E firms that are sometimes employed for quality assurance have the advantage that the design architect or engineer is also the inspector, assuring knowledge of contractual requirements. The designer is in a good position to check conformance, review changes, and establish acceptable alternates. A disadvantage from an owner's point of view is the potential conflict of interest of the designer approving his or her own work, which may lead to omissions and errors. However, the owner's design budgets for small projects will seldom accommodate independent third-party inspection services.

For mid-size commercial buildings—those in the $5–10 million dollar range, which constitute a substantial proportion of the private sector office and retail market—budgets are sufficiently large to accommodate special QA/QC personnel. The owner will often assign one or two staff members as the quality assurance observers. In some instances, such an observer is designated the "clerk of the works."[20]

As scale increases, large A/E, engineer-construction (E-C) and construction firms that work on large projects often have separate in-house QA/QC groups with independent reporting procedures and authority. The A/E, E-C and the constructor often designate a field quality assurance coordinator, and reporting to the coordinator may be one or more inspectors and a testing organization. The field coordinator reports to the home office on site problems such as owner changes, design changes or errors, construction errors, vendor changes or errors, and shipping damages. The home office audits the project for compliance with project procedures and checks that the field

[20] Clerk of the works is a British term for one who supervises the construction and keeps records of materials used. The clerk of the works may also be responsible for recording workmen's time. While the term is understood by the American building industry, Americans usually use the term project supervisor or representative. (Greater Phoenix Chapter of the National Association of Women in Construction, 1973. Construction Dictionary.)

quality assurance coordinator's records are clear and current. On projects with stringent time constraints, efforts will be made to assure that equipment is in compliance with specifications before it leaves the vendor's shop so that delays correcting problems in the field are avoided. Vendor pre-qualification is becoming more widespread.

LEGAL DISTINCTIONS BETWEEN FEDERAL AND PRIVATE PRACTICES

In addition to the applicability of local codes and municipal regulatory practices, other legal distinctions between public and private work influence quality management practices. The terms and conditions of the American Institute of Architects General Conditions of the Contract for Construction (AIA Document A201, 1987 Edition) illustrate typical industry practices in the private sector. The AIA also publishes related contract documents that define the architect's (Document B141) and the subcontractors' (Document A401) roles on the project. [21] The important legal feature of these documents is the formal assignment of responsibilities for QA activities to the design professional.

The term used in the AIA documents to describe the process during construction whereby the owner's interests are protected is "administration of the contract."[22] The architect oversees this process. The owner retains the architect to design the

[21] Published and copyrighted by the American Institute of Architects, these documents receive widespread use throughout the construction industry and are approved by the Associated General Contractors of America.

[22] In the AIA Contract Documents, "supervision" is a term used to describe the contractor's responsibilities on the construction project. Supervision connotes the contractor's management, direction, and control of the construction process; this is in contrast to the architect's visits to the site to observe the progress of construction and to evaluate the contractor's compliance with contract requirements.

project and to prepare the working drawings and specifications. After the contractor is hired, the architect "administers" the construction contract on behalf of the owner to determine if the contractor is fulfilling contract requirements and to advise the owner whether the contractor is performing the work properly.

The AIA contract requires the architect to visit the site "to become generally familiar with the progress and the quality of the completed work and to determine in general if the work is being performed in a manner indicating that (it) will be in accordance with the contract documents." The AIA General Conditions, however, make a clear distinction between the roles of the architect and contractor, and there is well-recognized language that states that the architect "is not responsible for construction means, methods, techniques, sequences or procedures, or for safety precautions and programs in connection with the work, since these are solely the contractor's responsibility . . . ." By maintaining this distinction in the contract between the roles of the architect and the contractor, the architect is able to concentrate on protecting the owner's interest.

Other major areas of responsibility for the architect spelled out in the AIA documents involve evaluating the contractor's applications for payment; rejecting work that does not conform to the contract requirements; taking appropriate action on shop drawings, samples and product data; preparing change orders; conducting inspections to determine the dates of substantial completion and final completion; and making the initial determination on claims by either the owner or the contractor.

In the AIA documents, the word "inspection" is used to describe the architect's services only in the context of determining substantial completion and final completion. At those two points in the construction process, the architect is contractually obligated to "inspect" (i.e., look very closely at) the work to determine compliance with contract requirements. The word "inspect" is used to connote the higher degree of care expected of the design professional at those two points in the process, in contrast to the more cursory evaluation that takes place during the site visits prior to the date of substantial completion.

COMMITTEE ASSESSMENT OF AGENCY PRACTICES

Overall the committee is impressed with the quality, dedication and success of every Federal Agency in performing its mission. Nevertheless the committee notes that there is room for improvement and is encouraged by the fact that the agencies are seeking ways to continually improve.

Agency quality management activities have been strongly questioned in some quarters. Evidence cited involves observed costs for correcting construction deficiencies. For example, one audit of selected DoD projects found that 82 percent of those projects contained . . . "defects that the construction contractors should have been required to fix either during construction or when it later became evident that the contractors had deviated from the plans and specifications." [23] Some critics argue that agencies depend too heavily on contractors' quality control systems for quality assurance on federal projects. These critics often call for increased inspection by agency staff or third-party contractors. While the committee does not share this view, we are convinced that agency practices can and should be improved, particularly with respect to the following observations:

1. Agency quality management practices vary too widely, especially when comparing DoD and DoE practices. Contractors performing work for more than one agency find themselves having to perform differently in different situations. This causes confusion, loss of efficiency, and an adverse effect on quality. One small, but important, improvement that should be made is to develop consistent definitions for quality and quality-related actions among all agencies.
2. Of all the agency practices reviewed, the committee considered the Corps of Engineers quality practices to be the most

[23] Department of Defense, December 31, 1984. Inspection Procedures and Value Engineering Design in the Real Property Construction Program, Report No 85-055, Audit Report of the Office of the Inspector General.

fully developed and effective, primarily because of the Corps' Construction Contractor Appraisal Support System (discussed in Chapter 3).[24] Other agencies would do well to emulate these practices.

3. The FAR and its amplifications by specific agencies are not as restrictive to improvements as the agencies believe. On the whole, the FAR is an excellent guide, and agencies can work effectively within its guidelines to take more innovative approaches than they are now executing. Some of these innovative approaches are discussed in the next chapter.

REFERENCES

Building Research Board, Committee on Construction Change Orders, 1986. Construction Contract Modifications: Comparing the Experiences of Federal Agencies and Other Owners, National Academy Press, Washington, DC.

Office of the Federal Register, 1985. Federal Acquisition Regulations System, Chapter 1, Parts 1 to 99, Code of Federal Regulations, 48CFR 1.000, U.S. Government Printing Office, Washington, DC.

Newman, Robert B., June 1989. Use of Consultants for Construction Engineering and Inspection, NCHRP Synthesis of Highway Practice 146, Transportation Research Board, Washington, DC.

[24] Otherwise, the Corps and NAVFAC follow similar procedures. See, for example, Contractor Quality Control (CQC) System, NFGS-01400 (February 1991), Department of the Navy; Contractor Inspection System, NFGS-01401 (February 1991), Department of the Navy; and Contractor Quality Control, CEGS 01440 (January 1991), Department of the Army.

3

EXTENSIONS AND ALTERNATIVES TO INSPECTION

Quality facilities—buildings and other construction that protect and enhance safety, productivity, and overall quality of life—are achieved through a complex interaction of many participants in the facilities development process. Inspection is one important and effective tool that serves the task of assuring quality, a tool that has evolved and continues to change in parallel with construction practices. Within the context of these practices, there are extensions and alternatives to inspection that federal agencies may use to improve their ability to achieve quality facilities. The committee drew on its assessment of agency and private practices and the knowledge of the committee's members to suggest what these extensions and alternatives are and how they can be used.

TOTAL QUALITY MANAGEMENT IN CONSTRUCTION

While they do not fault specifically the practices and achievements of any of the federal agencies' quality manage ment activities, the committee points out that many participants and observers of U. S. construction believe that the industry as a whole has serious problems. As discussed in Chapter 1, both the private and public sectors share these problems, experiencing serious cost growth due to a lack of quality. Industry groups such as the Business Roundtable, the Construction Industry Institute, and the ASCE have documented that the quality of the industry's work has declined in recent years and that the nation's productivity and international competitiveness have suffered. According to the director of the Construction Industry Institute, recent studies have shown that more than one-third of our current projects fail to meet budget objectives, a similar proportion finish behind schedule, and only about 80 percent meet technical objectives (Tucker, 1990). Industries cannot survive on such a low rate of satisfactory product delivery.

The industry has begun to work for change. A theme originated by U. S. consultants and educators W. E. Deming and J. M. Juran and adopted wholeheartedly in Japan—Total Quality Management (TQM)—has been gaining increasing popularity in the U. S. construction industry (The Quality Management Task Force, 1990). TQM is an organization-wide effort to improve performance that involves everyone and permeates every aspect of the organization to make quality a primary strategic objective. TQM is achieved through an integrated effort among personnel at all levels to increase customer satisfaction by continuously improving current performance.[25]

The quest for continuous improvement, one of Deming's "fourteen points," has in fact become a keystone of Japanese practice.[26] Staff at all levels are encouraged to remain vigilant, to bring opportunities for improvement to managers' attention. Improvements occur through a team effort to assess, adapt, and apply all suggestions for enhancement of product, process, or service.

[25] A brief discussion of the principles of TQM are included in Appendix D.
[26] The Japanese have a word, *keizen*, for this practice.

Then-Secretary of Defense Frank Carlucci stated the DoD's recognition of TOM in a 1988 memorandum. ". . . I am giving top priority to the DoD Total Quality Management (TQM) effort as the vehicle for attaining continuous quality improvement in our operations, and as a major strategy to meet the President's productivity objectives under Executive Order 12552." The committee believes that all federal agencies should implement TQM throughout the life cycle of their construction projects.

TEAMWORK AND QUALITY

One key aspect of TQM is teamwork, an integrated effort by all participants in the construction process to produce a quality building. A major conflict between current practice and TQM is the adversarial relationship among owners, designers, and constructors established by traditional inspection-based QA programs. This relationship—which can become especially severe if third-party professionals[27] are responsible for QA inspections—has the unfortunate consequence that participants become concerned primarily with avoiding blame when construction documents, constructed facilities, and owner's needs are poorly matched. These participants then feel little incentive to anticipate, prevent, and help to resolve such disputes when these mismatches arise.

Greater teamwork for most government agencies would mean an increasing role in the construction phase for the architect or engineer who designed the project. Rather than viewing this increased role as a substitute for the agency's staff who traditionally have performed construction inspection on federal projects, the A/E and agency staff would complement one another, the former focusing on the project's technical

[27] Such professionals could be members of the owner's organization, who were not involved in the building's design, or an A/E firm or other consultant providing QA services.

requirements and the latter responsible for administrative procedures. The details of staff roles should be responsive to characteristics of projects and agencies.

Teamwork extends to the construction contractor as well. Current laws and regulations governing federal agency construction make it difficult to reward good contractor performance with preferential or non-competitive award of future construction contracts. The reward of future work and lasting business relationship, common in the private sector, is an incentive for excellence that reinforces a quality contractor's own pride of workmanship. However, the government contractor must compete in the same way for new federal work, regardless of past quality or level of performance, and so needs only to perform adequately to prosper.

CONTRACTOR PERFORMANCE EVALUATION

While preferential treatment in bidding for future work is not readily available, federal agencies can develop mechanisms to strengthen construction contractors' incentives to deliver quality. Section 36.201 of the FAR requires agencies to prepare performance evaluation reports.[28] The DFARS also require that all DoD agencies forward these reports to a central data base maintained by Corps' North Pacific Division. This data base, called the Construction Contractor Appraisal Support System (CCASS), is the result of a July 1986 memorandum by the Chief of Engineers to the Under Secretary of the Army and has been on-line since July 1987. At the end of 1989, there were more than 13,000 reports on nearly 10,000 contractors in the system, with ratings assigned as outstanding, satisfactory, or unsatisfactory. (Of 13,251 final evaluations, 7.0 percent were outstanding, 84.2 percent satisfactory and 8.8 percent unsatisfactory.)

[28] Using Standard Form (SF) 1420.

In addition to the data base, the system includes periodic analysis of contractor performance and interim appraisals. Unsatisfactory appraisals are used to motivate the contractor's management to take corrective action.

The Corps has also established[29] a board to review CCASS data for initiation of debarment actions, and revised the EFARS to require contracting officers to make a determination on pursuit of debarment action for each final unsatisfactory performance rating. There have been three performance-based debarments and three more were under review in early 1990. Adoption of such a process, including coordinated data exchange, uniform policies, and consistent enforcement by all federal agencies could become a powerful tool for more effective quality management.

BIDDER PRE-QUALIFICATION

Even in the absence of past performance information, prequalifying contract bidders is a useful quality management tool. It should be more widely used. Criteria for pre-qualification of contractors can include requested references testifying to the bidders management quality and financial strength, visits to the contractors' recently completed projects, and contractors' presentations of their work. The pre-qualification procedure may continue after bids are received by holding interviews and additional reviews. This practice is sometimes criticized as "bid shopping"—inviting a contractor to underbid another's offer—but the committee finds that most contractors will accept this process as a reasonable business procedure.

[29] Under Engineer Regulation (ER) 15-1-29.

INCENTIVE CONTRACTING

Incentive contracting, widely used in the private sector, is a system whereby the contractor receives a bonus for performance above some pre-agreed base or norm, and is penalized for performance below that base. Such factors as cost, schedule, quality, safety, responsiveness, and management effectiveness may be negotiated as a basis for incentive payments, and amounts in the range of 0.33 percent to 1.0 percent of the construction cost are typical. In the committee's view, experience and research have shown that positive incentives for good performance are more useful than penalties for failure to meet targets.

For such incentives to work, the plan for their use must be simple. Reward should be commensurate with the risk the contractor is asked to accept, and all participants must be committed to the success of the incentive plan and continuously act in each other's best interest. Incentive awards therefore work best when they are passed along to subcontractors and craftsmen on the project. The plan must be such that if one party to the agreement wins, all win. The incentive payments should be viewed not as additional costs, but rather as part of the savings achieved by the owner when the contractor performs particularly well.

Use of as much quantitative measurement as is possible, consistent with making the plan simple, facilitates administration of the incentive contract. A team is formed to oversee evaluation and award of incentives, with equal representation by the contractor and owner. A higher level management team, also with equal representation of both parties, is formed to resolve disputes that may arise. In practice, the evaluation team typically reconciles differences rather than reporting to their superiors that they cannot arrive at a decision.

INTEGRATED INSPECTION PLANS

When construction is supervised by an organization other than the user, disagreements and user dissatisfaction can arise during construction because inspections for quality assurance do not adequately address the users concerns. The integrated inspection plan (1) has input from the design, construction, and inspection organizations, (2) documents their concurrence in the plan, and (3) contains mutually agreed upon criteria for acceptance or rejection of work. Although the construction contractor is not generally involved in formulation of the plan, it is important that the contractor both understand the contents of the plan and recognize that the engineering, construction, and inspection organizations concur with it.

An integrated inspection plan helps dispel a contractor's concerns about inspectors over-inspecting, an owner's concerns about the constructor's preferences for meeting schedule at the expense of doing the job right, and the designer's inclination to view a job as finished when the design is delivered to the owner. Potential sources of dispute surface during the review of the inspection plan and are resolved prior to the actual performance of work in the field. Like any other formal and consistent inspection plan, an integrated plan helps to assure the uniformity of inspections from one inspector to another and reduces the frequency of office consultations to review standards for acceptance of work.

COST AND SCHEDULE PROTECTION PLANS

The cost and schedule protection plan is a management tool intended to prevent problems that pose threats to cost or schedule due to the lack of conformance to required workmanship or material characteristics. Prior to construction, inspection personnel review construction schedules and specifications to identify those activities that can cause quality problems that may adversely affect the cost or schedule. These activities, termed "quality critical," typically have one or more of the characteristics listed in Table 3.1.

Quality-critical activities are included in the cost and schedule protection plan, which identifies the specific critical work manship, processes or materials and the hypothesized modes of failure which are the most probable sources of quality failure. The applicable criteria for judging the risk are then specified and the inspections and tests needed to manage this risk are scheduled. Finally, the plan designates the persons or organizations responsible for inspections, tests or other actions designed to monitor activity and provide early warning of impending problems.

TABLE 3.1 Indicators of quality critical activities (Source: committee experience)

Activities are likely to be *quality critical* if they are
-on the critical path or an accelerated schedule
-repetitive and a generic defect or fault would necessitate many repairs
-very labor-intensive to repair
-an intensive user of expensive or hard-to-get materials
-critical to facility operation
-likely to be inaccessible for repair
-historically a problem or source of high reject rates
-dependent on high skill levels or certifications
-dependent on special processes such as heat treating
-subject to approval by outside organizations
-complicated by specification, drawing or interface ambiguities
-a user of in-place storage, temporary construction shoring, tracing, or weather proofing that could lead to damage of expensive equipment and materials
-dependent on highly stressed and structurally significant components, high energy fluid systems, or other components whose failure would create life-threatening hazards to personnel

When conditions arise that may threaten the cost and schedule, the inspection organization immediately notifies the construction organization and recommends corrective actions and recurrence control. Timely detection of unacceptable conditions precludes further wasteful processing of defective work and greatly reduces the costs and delays associated with repairs.

Through early detection, a larger percentage of craft man-hours are spent on work that meets specifications the first time.

NEW TECHNOLOGY FOR INSPECTION

Emerging tools and procedures that utilize modern electronics, computers, and other technologies new to the field of construction quality control and assurance may in coming years revolutionize the industry. The task of inspection will be accomplished more quickly, using less labor, and yielding more reliable information about the quality of construction in-place. Federal agencies have much to gain from such improvements and can foster their development through research and demonstrations in the field.

For example, research on how the electrical and chemical properties of Portland cement concrete change during mixing and curing may lead to inspection methods that replace slump tests[30] at the time of placement and compressive strength testing of cylinders days after the concrete is placed. Transmission and resonance behavior of structural members, roof coverings, and wall sheathing materials (brick and other masonry, in particular) exposed to sonic and radio-frequency waves may become the basis for monitoring for voids and failures to achieve bonding at the interfaces between different materials.

Computers and communication technology are enhancing both constructors' and owners' abilities to review inspection data and monitor construction performance. A voice-mail system developed by the Army's Construction Engineering Research Laboratory permits inspectors to telephone their reports to a central office for typing and review, enhancing the speed and accuracy of inspection reporting. Laptop and hand-

[30] Slump is a measure of a physical characteristic of the still-plastic concrete mix, that indicates the workability of the mix and correlates broadly with later performance of the cured and finished concrete.

held computers are used in the field to facilitate access to design data and previous inspection information. Development of expert systems software will supplement the inspector's judgment in quality control.

Already in use are electronic surveying and leveling devices that have improved dramatically the dimensional accuracy of construction. These devices are also yielding benefits in faster construction because less time is required to tailor components on site to fit idiosyncratic results of preceding stages of construction. Such results—improved accuracy and construction time—assure true and unambiguous improvements in construction quality.

QUALITY MEASUREMENT SYSTEMS

The Construction Industry Institute has developed a Quality Performance Management System (QPMS), which classifies and tracks the costs of quality during the design and construction of projects.[31] This management tool is designed to determine what quality management activities and deviation categories are involved in a project and to ascertain when (i.e., during which project phases) the quality management activities and deviation costs occur. The manager is helped to discover why the deviations occur and to learn how the rework relates to the quality management. The tool thus can provide cost-of-quality information to establish baselines and identify opportunities for improvement, without providing either too much or too little detail.

Appendix E summarizes principal elements of the Quality Performance Management System.

[31] The Quality Management Task Force, February, 1990. The Quality Performance Management System: A Blueprint for Implementation, Publication 10-3, The Construction Industry Institute, The University of Texas at Austin.

REFERENCES

The Quality Management Task Force, May, 1990. Total Quality Management: The Competitive Edge, CII Publication 10-4, The Construction Industry Institute, The University of Texas at Austin.

Tucker, Richard L., May 1990. "The Big 'Q'," "Perspectives," The Construction Specifier, Volume 43, Number 5, 151–152.

4

CONCLUSIONS AND RECOMMENDATIONS

This study began with a questioning of procedures used by federal agencies for inspection during construction of their facilities. While the committee found that the inspection strategies used by the agencies, though varied, are adequate to provide a reasonable assurance that the specified requirements are being met, improvements can and should be made. Furthermore, as the preceding chapters have discussed, the committee felt this focus on inspection is too narrow to effectively address the real issue of assuring quality in federal construction. The committee's deliberations, conclusions, and recommendations—reflecting their judgment and experience—spanned this broader issue.

GETTING QUALITY IN FEDERAL FACILITIES CONSTRUCTION

"Quality," within the limited context of construction, is defined as conformance to adequately developed requirements.

Current practices for establishing, stating, and communicating requirements and for determining that construction has indeed met these requirements offer many opportunities for mistakes, misunderstandings, and oversights. An essential precondition for assuring construction quality is getting the requirements right, and the committee recommends that *federal agencies should continue working to improve their ability to develop facility programs, plans, budgets, guide criteria, design drawings, and specifications that convey the appropriate requirements in a clear manner to the constructor.*

Agency personnel, private architects and engineers employed by agencies to plan and design specific facilities, and constructors employed under contract to build these facilities all have roles to play in getting the requirements right. Quality is more likely to be assured when these parties work cooperatively toward the common goal of delivering a facility that meets the agency's needs. *Agencies should avoid adversarial design and construction management practices and adopt defined programs to foster teamwork among users, design and construction managers, designers, and constructors.* The TQM philosophy is a worthy basis for formulating these programs (See Appendix D).

THE DESIGNER'S ROLE

Architects and engineers who plan and design federal facilities play a key role in determining the quality of these facilities. *Both the agencies and their designers should work to assure that the drawings and specifications that present requirements to the constructor are a complete and clear statement of what the owner and user expect in the facility.* In those cases where a construction agency other than the user is responsible for administration of the building process, all of these parties must work even harder. The TQM philosophy is an appropriate basis for this effort as well.

THE ROLE OF INSPECTION

Inspection, only one of several elements in an effective quality management program, is nevertheless an important means for controlling conformance to requirements and an essential part of any quality management program. However, the value of inspection has limits and over-inspection wastes agency resources, adds to the cost of construction, and establishes unproductive adversarial relationships among owners, designers, and constructors. *Agencies should avoid excessive controlled and superfluous discretionary inspections.* A systematic review by each agency of its inspection practices, conducted within the context of cost and schedule protection planning, can reveal when inspections are being called for out of proportion to the importance of the inspected items to overall quality.

COORDINATED AGENCY QUALITY MANAGEMENT PROGRAMS

The currently independent quality management programs of individual agencies can be made more effective by joint action. *Agencies should join in* specific programs to share information and centralize selected inspection activities. Broad participation in a contractor performance monitoring system (such as the Army's Construction Contractor Appraisal Support System) would increase the importance of the system to contractors and enhance its contribution to quality of agency construction.

Establishment of a cadre of trained inspection professionals, based at federal district or area office levels and serving all agencies constructing facilities within that district or area, would enhance the government's ability to maintain adequate personnel and provide these professionals with greater career opportunities within their field of expertise. These centralized quality management resources can be expanded to include data analysis support and purchasing of testing services and training of managers to be open and responsive to workers' proposals.

However, the goal of total quality management and the search for continuous improvement must penetrate all parts of the organization.

The committee further recommends all federal agencies *adopt consistent quality definitions (Appendix B) and standardize*, insofar as possible, *agency practices*, patterned after the Corps of Engineers.

EFFECTIVE INSPECTION

The resources available for inspection must be deployed effectively and will be most productive when all parties to the facilities development process accept the value and relevance of inspections. To assure this effective deployment and acceptance, *agencies should develop integrated inspection plans for their construction projects*. These plans, prepared jointly by the agencies and their design consultants, should be reviewed with the construction contractor and accepted prior to commencing construction, perhaps at the pre-bid conference, as part of the contract negotiations, and at the preconstruction conference. When the user agency and construction agency are different, both agencies should be involved in developing the plan.

The federal government as a whole is the nation's principal purchaser of construction services and can be a powerful force for advancing the state of the art in construction quality management. *Agencies should fund the research and demonstration activities required to develop new inspection and other quality assurance technologies.*

QUALITY MEASUREMENT SYSTEMS

Federal agencies should adopt systems for measuring their quality management efforts and relate these efforts to the costs associated with doing things over. The Construction Industry Institute's Quality Performance Management System provides a simple, cost-effective, management tool to inform agency personnel where improvements can be made. Coupled with a TQM focus, quantitative quality measurements will help foster change and provide the information needed for positive improvements in quality to be effected.

EFFECTIVE QUALITY MANAGEMENT

Agencies seek quality facilities that enhance safety, productivity, and overall quality of life. This quality is assured only when there is a commitment to quality throughout planning, design, and construction. This commitment must be extended to operations and maintenance of facilities as well. *Senior agency administrators should assert this commitment and establish definite programs for making the commitment effective and lasting.*

This recommendation applies not only to construction agencies, but to users as well. Satisfaction of users' needs is the source of requirements to be met in construction and the ultimate measure of quality. Effective construction quality management requires that the user's interests be reflected.

ASSURING QUALITY IN CONSTRUCTION

Quality in construction is a product of the complex interaction of many participants in the facilities development process. The committee's recommendations, aimed primarily at agency managers, address design and construction professionals, educators, and policy makers as well. Quality in construction is assured only when there is a commitment to quality throughout all stages of the facility service lifetime, from initial conception through operations and maintenance. Quality facilities that meet and exceed expectations—enhancing our safety, productivity, and overall quality of life—are the result of such commitment and the real goal of the committee's work.

APPENDIX A

BIOGRAPHICAL SKETCHES OF COMMITTEE MEMBERS AND STAFF

WILLIAM B. LEDBETTER, PE (*Chairman*), is active as a consultant and researcher and formerly Professor of Civil Engineering at Clemson University. He earned a B.S. degree from Texas A&M University in 1956, and the doctorate degree in Civil Engineering from the University of Texas in 1964. He is an expert in the quality management aspects of design and construction of major industrial projects, and through his research and projects has authored many technical publications. He has received several national awards from professional organizations and served on national committees, including chairing the Quality Assurance Task Force of the Business Roundtable's Construction Industry Cost Effectiveness Project.

MERRILL R. COTTEN is Chief of Construction Management for the Chicago Office of De Leuw, Cather & Company. He received a B.S.C.E. in 1951 from the University of Missouri School of Mines & Metallurgy. Mr. Cotten has over 35 years experience in management of major construction projects. He is currently supervising and coordinating all aspects of construction management, including developing properties. He is responsible for project control, scheduling, procurement, preparing cost estimates and specifications, and field supervision for domestic projects.

FLETCHER H. GRIFFIS is Professor of Civil Engineering at Columbia University, heading the Construction Engineering Program and serving as Director of the Center for Infrastructure Studies. He received a B.S. degree from the U.S. Military Academy in 1960, an M.S. in Construction Management in 1965, Ph.D. in Civil Engineering in 1971, and M.S. in Industrial Engineering in 1972, from Oklahoma State. Prior to his retirement from the U.S. Army Corps of Engineers, he served as Commander and District Engineer of the New York District, responsible for all Army and Air Force construction from Southern New Jersey to Maine, and in Greenland and Labrador. His current interests include the applications of expert systems and 3-D models to construction engineering, modular housing design, pavement systems and facilities management.

C. RAYMOND HAYS is Director Safety/Quality for RUST International. He received his B.S. in Civil Engineering in 1953 and M.S. in Engineering in 1972. He is responsible for developing and implementing quality control programs for construction projects. With a background in design, Mr. Hays has spent fifteen years in construction related activities.

ARTHUR T. KORNBLUT is a principal in the law firm of Kornblut & Sokolove. He is a member of the bar in D.C., Maryland, and Ohio, holds an architect's license, and an NCARB Certificate. He specializes in construction law, authoring the "Legal Perspectives" column in *Architectural Record*, and serving on the panel of arbitrators for the American Arbitration Association. Mr. Kornblut was Chairman of the American Bar Association Forum Committee on the Construction Industry from 1985 to 1987, and served on the BRB Committee on Post-Occupancy Evaluation Methodology.

JOHN WOODLAND MORRIS is engineer advisor to Zorc, Rissetto, Weaver & Rosen, and an adjunct professor of construction management at the University of Maryland. He was formerly Chief of Engineers, U.S. Army Corps of Engineers, Executive Director for International Operations for Royal Volker Stevin N.V., and Chairman/CEO of Planning Research Corp. Engineer Group. He earned a B.S. in Civil Engineering from the U.S. Military Academy in 1943 and an M.S. from the University of Iowa in 1948. He is an expert in construction management and has received numerous awards and honors from professional societies and government agencies for his construction and management accomplishments, including a Presidential Citation for Management when Chief of Engineers, and Construction Man of the year (1977) from *Engineering News Record* and the Pladium Medal sponsored by the Audubon Society. He is a member of the National Academy of Engineering.

DONALD W. PFEIFER is Vice President of Wiss, Janney, Elstner Associates, Inc. He received a B.S. in Civil Engineering in 1958, and an M.S. in Theoretical and Applied Mechanics in 1960, from the University of Illinois in Urbana, and is a Registered Structural Engineer in Illinois. He has carried out a wide range of experimental research investigations concerning structural lightweight aggregate concretes, precast panels, architectural concrete, and laboratory and field instrumentation studies on reinforced concrete buildings. The results of his work have been published extensively. He specializes in investigation and repair of deteriorated concrete structures.

ROBERT ULREY is Manager of Technical Services for the Houston Lighting & Power Company, responsible for the inspection, testing, non-destructive examination, and materials engineering for fossil-fired power plants and associated projects. He has extensive experience in quality assurance, management, and codes compliance, related to missiles, space vehicles, ground support equipment, and test facilities.

DAVID J. WICKERSHEIMER, is Professor of Architecture at the University of Illinois, Champaign, and President of Wickersheimer Engineers. He earned the B.Arch. degree in 1967 and M.S.Arch. in 1969 from the University of Illinois. He is a registered professional engineer and architect with special expertise in the prevention and investigation of building failures and the practices of building inspection.

PAUL ZIA is Distinguished University Professor of Civil Engineering at North Carolina State University. He received a B.S. in Civil Engineering from National Chiao Tung University, China, in 1949, an M.S. in Civil Engineering from the University of Washington in 1951, and the Ph.D. in Civil Engineering from the University of Florida in 1960. His areas of expertise include innovative design, basic and applied research, failure investigation and strength evaluation of reinforced and prestressed concrete structures, and properties and applications of high performance concrete. He is the 1989 President of the American Concrete Institute, an honorary member of ASCE and a member of the National Academy of Engineering.

STAFF

ANDREW C. LEMER, Director, is an engineer-economist and planner. Formerly division vice president with PRC Engineering, Inc., Dr. Lemer is founder and president of the MATRIX Group, Inc., and has written widely on matters of building economics and development policy, often in conjunction with his work on major projects in the United States and overseas. He received his S.B., S.M., and Ph.D. degrees in civil engineering from the Massachusetts Institute of Technology and is a member of the American Institute of Certified Planners, the American Society of Civil Engineers, the Urban Land Institute, and the American Macroengineering Society.

PETER H. SMEALLIE, Director of the Geotechnical Board, and formerly Senior Program Officer, and Executive Secretary of the Public Facilities Council maintains a working arrangement with the BRB on programs of mutual interest. Prior to his work with the NAS-NRC, he was Vice President of Thomas Vonier Associates, an architecture and consulting firm, and was a program director with the American Institute of Architects Research Corporation. He has a degree in Urban Studies from St. Lawrence University.

APPENDIX B

GLOSSARY OF TERMS RELATED TO CONSTRUCTION QUALITY

There is considerable disagreement and in some cases confusion about terms used in connection with quality assurance and control in construction. The following definitions were adopted by the committee, primarily from federal agency and Construction Industry Institute sources.

ACCEPTANCE CRITERIA. Specified limits placed on characteristics of a product, process, or service defined by codes, standards, or other requirement documents.

APPRAISAL. Activities employed to determine whether a product, process or service conforms to established requirements, including: design review, specification review, other documentation review, constructibility review, materials inspection/tests, personnel testing, quality status documentation, and post project reviews.

AUDIT. A formal, independent examination with intent to verify conformance with established requirements. An audit does not include surveillance or inspection for the purpose of process control or product acceptance.

CHANGE. A directed action altering the currently established requirements. Changes may address design, fabrication, construction. and materially affect the approved requirements, the basis of design, the existing scope of the contract plans and specifications, or operating capability of the facility.

CORRECTIVE ACTION. Measures taken to rectify conditions adverse to quality, and where necessary, to preclude repetition. Corrective action includes rework for non-conformance and deviations.

COST OF QUALITY. The cost associated with quality management activities (prevention and appraisal) plus the cost associated with deviations.

CRITICALITY. A measure of the significance or impact of failure of a product, process or service to meet established requirements.

DEFECT. A deviation with a severity sufficient to require corrective action.

DEVIATION. A departure from established requirements. A deviation may be classified as an improvement, an imperfection, non-conformance, or defect, based on its severity.

DEVIATION COSTS. The sum of those costs, including consequential costs such as schedule impact, associated with the rejection or rework of a product, process or service due to a deviation.

ERROR. Any item or activity in a system that is performed incorrectly, resulting in a deviation, e.g., design error, fabrication error, construction error, etc. An error requires an evaluation to determine what corrective action is necessary.

IMPERFECTION. A deviation which does not affect the use or performance of the product, process or service. In practice, imperfections are deviations that are accepted as is.

INSPECTION. An activity involving the appraisal of a process, product or service to ascertain if it conforms to the established requirements. An activity which the contractor has an obligation and the government has a right to perform.

ITEM. An all-inclusive term used in requirements documents in place of any of the following: appurtenance, assembly, component, equipment, material, module, part, structure, subassembly, subsystem, system, or unit.

NON-CONFORMANCE. A deviation that occurs with a severity sufficient to consider rejection of the product, process or service. In some situations the product, process or service may be accepted as is; in other situations it will require corrective action.

OMISSION. Any part of a system, including design, construction and fabrication, that has been left out resulting in a deviation. An omission requires an evaluation to determine what corrective action is necessary.

PREVENTION. Activities employed to avoid deviations, including such activities as: quality systems development, quality program development, feasibility studies, contractor/subcontractor evaluation, quality orientation activities and certification/qualification.

PROJECT. All those elements and activities associated with a facility from initial concept to final disposition.

PROJECT ELEMENTS. The major phases of a project, including preplanning, design, procurement, construction, start-up, operation and final disposition.

QUALITY. In construction, conformance to established and adequately developed requirements.

QUALITY ACTIVITIES. Those activities in a project directly associated with problem prevention and appraisal.

QUALITY ASSURANCE. All those planned or systematic actions undertaken to provide adequate confidence that a product, process or service will conform to established requirements.

QUALITY ASSURANCE INSPECTOR. The person who performs inspection for quality assurance.

QUALITY CONTROL. Inspection, test, evaluation or other actions to verify that a product, process or service conforms to established requirements.

QUALITY CONTROL INSPECTOR. The person who performs inspection for quality control and, after appraisal, recommends corrective action.

QUALITY MANAGEMENT. The process of optimization of the use of resources for quality activities; includes prevention and appraisal activities.

QUALITY MANAGEMENT COSTS. The sum of those costs associated with prevention.

QUALITY PERFORMANCE MANAGEMENT SYSTEM. A management tool providing data for the quantitative analysis of certain quality-related aspects of projects by systematically collecting and classifying quality management costs.

REQUIREMENT. A contractually established characteristic of a product, process or service. A characteristic is a physical or chemical property, a dimension, a temperature, a pressure, or any other specification used to define the nature of a product, process or service.

SURVEILLANCE. The act of monitoring or observing to verify whether an item or activity conforms to established requirements.

VERIFICATION. The act of reviewing, inspecting, testing, checking, auditing, or otherwise determining and documenting whether items, processes, services, or documents conform to established requirements.

APPENDIX C

FEDERAL ACQUISITION REGULATION AND CONSTRUCTION

The Federal Acquisition Regulation presents an extensive set of rules, policies, and procedures that govern government agency procurement of supplies and services. The committee identified the 36 sections cited in the following listing as particularly relevant to procurement of construction and administration of construction contracts.

1. Priorities and Responsibilities (Subpart 1.6).
2. Full and Open Competition (Subpart 6.1).
3. Sealed Bidding and Competitive Proposals (Subpart 6.4).
4. Acquisition Plans (Subpart 7.1).
5. Responsible Prospective Contractors (Subpart 9.1).
6. Delivery or Performance Schedules (Subpart 12.1).
7. Liquidated Damages (Subpart 12.2.).
8. Use of Sealed Bidding (Subpart 14.1).
9. General Requirements for Negotiation (Subpart 15.1).
10. Small Business and Small Disadvantaged Business Concerns (Part 19).

11. Certificates of Competency and Determinations of Eligibility (Subpart 19.6).
12. Labor Surplus Area Concerns (Part 20).
13. Application of Labor Laws to Government Acquisitions (Part 22).
14. Environment, Conservation and Occupational Safety (Part 23).
15. Drug Free Workforce (Subpart 23.5).
16. Buy American Act (Subpart 25.2).
17. Bonds (Subpart 28.1).
18. Sureties (Subpart 28.2).
19. Contract Cost Principles and Procedures, Construction and Architect-Engineer Contracts (Section 31.105).
20. Prompt Payment (Subpart 32.9).
21. Disputes and Appeals (Subpart 33.2).
22. Special Aspects of Contracting for Construction (Subpart 36.2).
23. Special Aspects of Sealed Bidding in Construction Contracting (Subpart 36.3).
24. Special Procedures for Negotiation of Construction Contract (Subpart 36.4).
25. Contract Clauses (Subpart 36.5).
26. Performance Evaluation of Construction Contractors (Paragraph 36.701(e)).
27. Contract Administration Office Functions (Subpart 42.3).
28. Contract Modifications (Part 43).
29. Subcontracting Policies and Procedures (Part 44).
30. Providing Government Property to Contractors (Subpart 45.3).
31. Quality Assurance (Part 46).
32. Value Engineering, Clause for Construction Contracts (Section 48.202).
33. Termination of Contracts (Part 49).
34. Contract Clauses (Part 52).
35. Inspection of Construction (Subsection 52.246-12).
36. Standard Form 1420 (Subsection 53.301-1420).

APPENDIX D

TOTAL QUALITY MANAGEMENT

Total quality management (TQM) is a complete management philosophy that permeates every aspect of a company and places quality as a strategic issue. It is accomplished through an integrated effort between all levels of a company to increase customer satisfaction by continuously improving current performance. Developed in a manufacturing mass production setting, TQM is an effective, comprehensive management technique that has proven successful both overseas and in the U.S., in services and in construction,[32] notwithstanding the fact that construction is a one-time process. Japanese construction companies began implementing TQM during the 1970s, and the industry in that nation has embraced the TQM concepts. Since the mid-1970s, three Japanese contractors have been awarded the coveted Deming Prize for quality improvement.[33]

[32] The Quality Management Task Force, op. cit.

[33] Deming, W.E., 1986. Out of the Crisis, MIT CAES, Cambridge.

Every party in a production process is seen to have three roles: supplier, processor, and customer. Juran, a principal proponent of the TQM approach, defines this as the "triple role" concept.[34] This concept can be illustrated for construction with the architect/engineer (A/E) being a customer of the owner, a processor of the design, and a supplier of plans and specifications to the contractor. The contractor is a customer for the A/E's plans and specifications, a processor of the construction, and a supplier of the completed structure to the owner.

Customer satisfaction is a key goal, whether customers are internal or external to the company. For engineering, the products are plans and specifications and the customers are the construction organization and the owner. For construction, the product is the completed facility and the customer is the final user of the facility. In design and construction organizations, internal customers receive products and information from other groups or individuals within their organization. Satisfying the needs of these internal customers is an essential part of the process of supplying the ultimate external customer with a quality product.

Under TQM, management has two primary functions directed toward continual improvement of the production process and subsequent increase in customer satisfaction. The first step is maintaining and incrementally improving current methods and procedures through process control. Then, efforts are turned toward achieving, through innovation, major technological advances in the production, i.e., engineering and construction, processes.

Most quality assurance and control experts define seven fundamental elements of TQM:[35]

[34] Juran, J.M., 1988. Juran on Planning for Quality, The Free Press, New York.

[35] Matthews, M.F., and J. L. Burati, July, 1989. Quality Management Organizations and Techniques, A Report to the Construction Industry Institute, CII Source Document No. 29, The University of Texas at Austin.

- Management Commitment and Leadership
- Training
- Teamwork
- Statistical Methods
- Cost of Quality
- Supplier Involvement
- Customer Service.

TQM is felt by some people in the industry to be an essential element in recovery of U. S. construction productivity and competitiveness in international markets. According to the Construction Industry Institute, companies which do not implement Total Quality Management will not be competitive in the national and international market within the next 5–10 years.[36]

[36] The Quality Management Task Force, op cit.

APPENDIX E

QUALITY PERFORMANCE MEASUREMENT

by William B. Ledbetter

To utilize the Quality Performance Management System, the project management team should divide a project into its major phases (e.g. design, construction, and start-up) and its major disciplines of work (Figure E1). Whenever possible, the disciplines should include those needed for the constructed product, but the number of disciplines should be kept to a minimum. It is very important that the discipline breakdown be consistent throughout ALL phases of the project. For example, an item of work during the design phase which is classified as CIVIL must not be classified as STRUCTURAL during the construction phase. Unless this is rigidly followed, the resulting information will be of little use.

Once a project is divided into its major phases and disciplines, the following three types of costs must be captured:

1. The normal cost of performing work (which is the productive work).
2. The quality management costs (by major activity).
3. The cost of rework (by primary CAUSE).

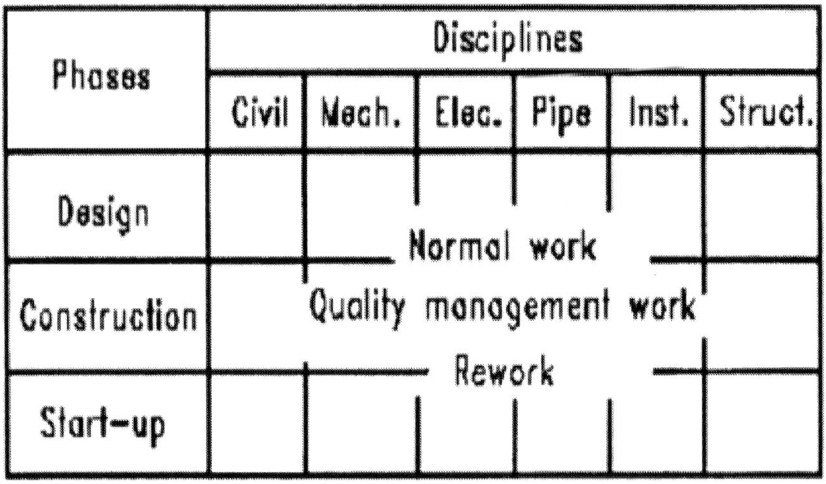

Figure E1
Interrelationships Between Project Phases and Major Disciplines

Most organizations use only a handful of activities as quality management, ranging from design and constructibility reviews to inspections and tests. They include both prevention and appraisal activities. Each organization should establish the specific quality management activities deemed necessary and develop clear definitions for them. The Construction Industry Institute (CII) lists 15.[37]

Rework is categorized by major cause. The cause is coupled with the time of detection (phase) for management action. The CII identifies a total of 26 possible deviation categories (marked "x" in Table E1).[38]

[37] The Quality Management Task Force, May, 1989. Measuring the Cost of Quality in Design and Construction, CII Publication 10-2, The Construction Industry Institute, The University of Texas at Austin.

[38] The Quality Management Task Force, February, 1990. The Quality Performance Management System: A Blueprint for Implementation, CII Publication 10-3, The Construction Industry Institute, The University of Texas at Austin.

TABLE E1 Major Causes of Rework, by Phase

Primary Cause	When Detected (Phase)			
(Party and Type)	Design	Procurement	Construction	Start-Up
Owner Change	x	x	x	x
Designer Error/Omission		x	x	x
Designer Change	x	x	x	x
Vendor/Error/Omission	x	x	x	x
Vendor Change	x	x	x	x
Construction/Error Omission		x	x	
Constructor Change		x	x	
Transporter Error		x	x	x

Organizations wishing to implement QPMS should perform the following steps:

1. Acquaint all personnel with the purpose and scope of the QPMS. This is a very important step! In many organizations there is a "cultural" bias against this type of effort that must be overcome if QPMS is to be successful. The organization will achieve the desired result only by assuring everyone that the results are only going to be used to help an organization improve the quality of its operations and the products and services it provides to its customers, thereby improving the organization's competitive position. Furthermore, the organization must assure personnel that the system will NOT be used as a tool for punishment.
2. Examine specific organizational needs for the data. Each organization must adapt QPMS to its specific needs and procedures. The organization decides which project phases are to be tracked and what major disciplines are involved. It should define disciplines clearly for the entire project and use IDENTICAL DISCIPLINE BREAKDOWNS FOR ALL PHASES.

3. Decide which quality management activities and rework causes should be tracked. Each organization and project is unique. Each organization must identify which quality management activities and rework causes are germane to most of the projects that the organization executes and define them clearly.
4. Modify the existing accounting system or set up an independent system to implement the QPMS. Some accounting systems are quite flexible and can be adapted to include the QPMS categories. Others will not accommodate the system. If the existing system is not adaptable, it is relatively easy to develop a stand-alone coding system which can be used in conjunction with a personal computer to capture, analyze, and report quality-related data with minimum effort.
5. Develop forms to capture relevant quality-related data. The organization may use its existing forms, if they track the quality-related data required by QPMS, or develop new forms. Some quality costs, such as those in Table E2, require frequent, even daily, tracking. Fortunately, continual record-keeping of such information is quite common in the construction industry and should represent no great additional burden. Some quality costs may be too small to justify tracking (e.g., materials used in design).

TABLE E2 Informational Requirements for Capturing Cost of Quality Data

Personnel times/costs
By salary or wage scale
By quality management activity or rework category
By discipline and phase
Equipment usage/costs
By hourly or daily rate
By quality management activity or rework category
By discipline and phase
Materials usage/costs
By quality management activity or rework category
By discipline and phase

6. Train the appropriate personnel in the procedures necessary to gather the correct information. Quality begins with training and ends with training. Correct information will only be obtained by adequate training of appropriate personnel in the QPMS information requirements.
7. Monitor and modify the organization-tailored system as necessary to capture and report the data in meaningful terms for management analysis and action. The costs of quality should be analyzed for each discipline, each phase of the project, and finally for the project as a whole. Since project cost breakdowns are typically provided at regular intervals, beginning with the estimate and ending with a wrap-up, this information can be integrated with the typical cost information to monitor quality performance and provide for rapid management action.

USE OF THE INFORMATION

Analysis of the cost-of-quality information on industrial projects from design through start-up indicates:

1. The information forms baselines for future project comparisons of performance.
2. Where the largest cost of quality expenditures occur.
3. The root causes of rework.

This knowledge provides the basis for decreasing the total costs for quality through optimization of quality management efforts leading to a minimization of rework.

However, organizations should be aware that, initially, both the perceived and actual cost of quality will increase (Figure E2). Increased awareness and emphasis on quality usually means more quality costs are discovered and documented. Once implemented, the QPMS will offer guidance on what improvements can be made. Implementation of these improvements will help improve the processes used to achieve quality and reduce the overall cost of quality!

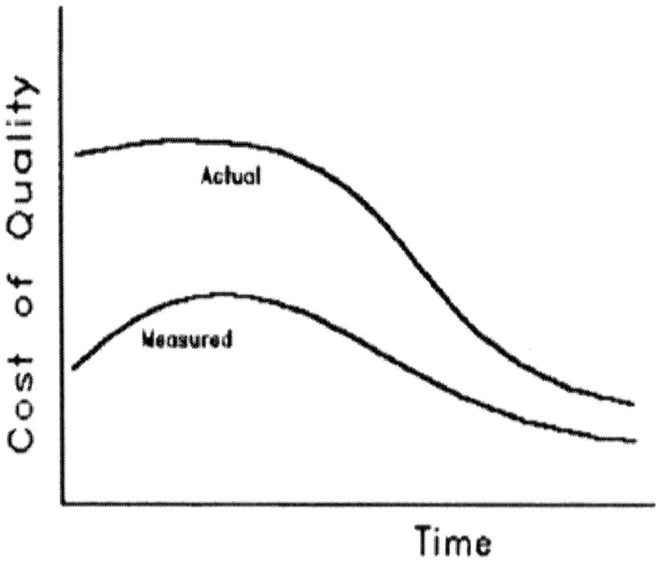

Figure E2
Measured and Actual Cost of Quality with Time for an Organization Implementing the Quality Performance Management System

The public sector of the construction industry can learn from the experience gained in the industrial sector. Although not an alternative to inspection, implementing the QPMS will provide managers with vital information on where improvements can be made, allowing them to optimize scarce inspection resources and achieve quality in a more cost-effective manner.